S0-FQY-057

KING COBRA

AND OTHER SENSATIONAL SERPENTS!

Written by Michaela Weglinski
Designed by Daniel Jankowski and Bill Henderson

All rights reserved. Published by Tangerine Press, an imprint of Scholastic Inc., *Publishers since 1920*. SCHOLASTIC, TANGERINE PRESS, and associated logos are trademarks and/or registered trademarks of Scholastic Inc.

The publisher does not have any control over and does not assume any responsibility for author or third-party websites or their content.

No part of this publication may be reproduced, stored in a retrieval system, or transmitted in any form or by any means, electronic, mechanical, photocopying, recording, or otherwise, without written permission from the publisher. For information regarding permission, write to Scholastic Inc., Attention: Permissions Department, 557 Broadway, New York, NY 10012.

Copyright © 2022 Scholastic Inc.

Tangerine Press
an imprint of
SCHOLASTIC
scholastic.com

10 9 8 7 6 5 4 3 2

ISBN: 978-1-338-81814-7
Printed in Guangzhou, China

Table of Contents

This book is about to take you around the world to visit some of the deadliest snakes on the planet.

Some snakes are mostly harmless to people, such as garter snakes, rat snakes, and green snakes that can be found in fields and around ponds and don't attack people unless they are bothered. But this book is not about them—oh no. While all snakes prey on living creatures, this book is about the specialized hunters with venomous fangs and the ability to constrict prey. You'll find out about cobras and vipers, which are deadly snakes that pack strong enough venom to kill animals—and even people— with one bite. You'll also learn about constrictors like pythons and boas— the predators that slowly squeeze the life out of their prey.

Within this book, you'll find some of the most feared, terrifying, and awesome creatures in the animal kingdom. They come in all colors, patterns, and sizes, and can be found slithering across land, in water, and even in the air (not kidding!), but they are all fierce and fantastic reptiles.

RULER OF SNAKES:

The Fearsome King Cobra

The king cobra is the longest and largest venomous snake in the world.

These creatures are called king cobras because they kill and eat other snakes, including cobras. Its genus name, or the first part of its name in scientific terms, is *Ophiophagus* (oh-fee-OFF-ah-gus), which means "snake eater"! King cobras can also rise up and look an adult human straight in the eye.

King Cobra

If this snake is disturbed, the king cobra can "stand up," with a third of its body off the ground, flare out its hood, and slither toward its enemy. Its scales are yellow, green, brown, or black with white or yellow markings.

These snakes are found in northern India, southern China, in the Malay Peninsula in Southeast Asia, and the Philippines. Adult king cobras are usually 10–12 ft. (3–3.6 m) long, but they can grow up to 18 ft. (5.4 m). It preys mostly on other snakes, but it will also feed on lizards, eggs, and small mammals.

King cobra compared to a 6 ft. (1.8 m) tall person.

HUNTING HABITS

While a king cobra can rise up to face its foes to protect itself and scare off enemies, striking a pose is not its only special ability. Here are a few things that make these skilled hunters so deadly!

Cobra Venom Packs a Punch

If the king cobra's warning signs don't work, it defends itself with its deadly bite. This snake's fangs are almost 1/2 in. (12.7 mm) long, and are venomous. That means it has special saliva, or spit, called *venom* that gets injected through its fangs like a needle when it bites its prey. The cobra's venom can kill a human in fifteen minutes, or a full-grown elephant in a matter of hours!

Bites Don't Bother This King

The king cobra hunts down and eats other snakes. But what happens if one of its snake snacks fights back? As it turns out, the venom of most other snakes doesn't affect the king cobra. Scientists found out that over time, the cobras were not hurt or killed by the venom of the other snakes they hunted. This happened often enough that the ability became part of their DNA, or genetic code.

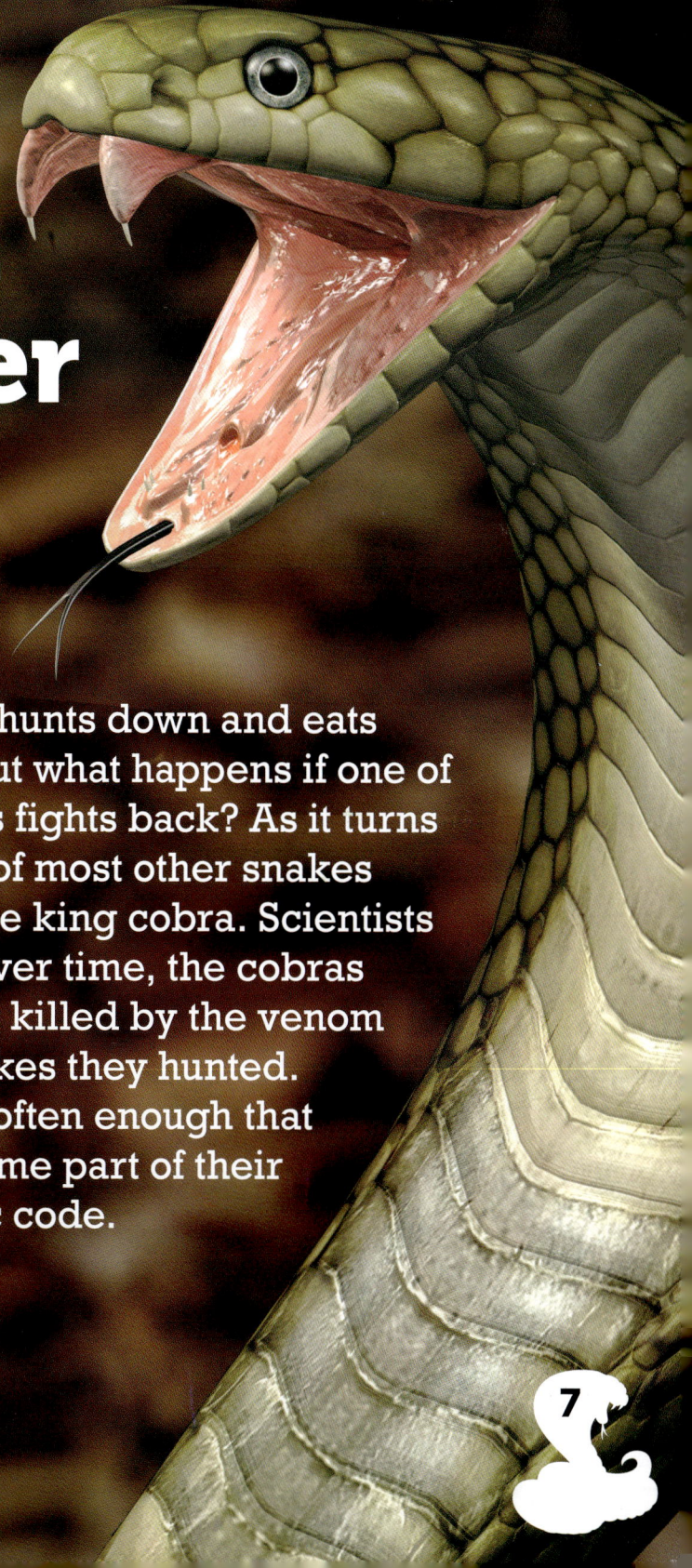

The Elapid Family Tree

King cobra may hold its reign over other venomous snakes, but its relatives in the Elapidae (uh-LA-pa-dee) family are nothing to sneeze at. Found in warm areas around the world, elapid snakes have fixed fangs that they use to deliver deadly venom. Here are just a few members of this fierce family!

Eastern Coral Snake

This colorful snake comes with its own warning sign. There's a famous rhyme that people use in the United States to tell the eastern coral snake apart from its less dangerous relatives: "Red touches yellow, kills a fellow; red touches black, friend of Jack." While you should never get close to a snake that you see in the wild, this rhyme is a great way to remember which red, yellow, and black snake found in North America has deadly venom. Strangely enough, this same rhyme does not work for many of the coral snakes in Central and South America!

Black Mamba

Extremely fast, deadly, and aggressive when disturbed: all of these things make the black mamba deeply feared. Despite its name, the snake is actually olive, brownish, or gray. But the inside of the snake's mouth is a dark blue or "inky" black color, which the snake shows off to scare enemies away. If you're close enough to a black mamba to see the inside of its mouth, you might be too late—the venom from its bite can kill a human within twenty minutes.

Indian Cobra

This elegant cobra often has markings on its hood that can look a bit like a pair of round glasses. This snake also belongs to a group called the "big four"—one of four types of snakes that are to blame for most of the worst snakebites in South Asia (including India).

Common Krait

The common krait is another member of the "big four" species that are to blame for most of the deadliest snakebite attacks in South Asia. This snake is also nocturnal, which means that it hides during the day and comes out at night to hunt. Unfortunately, the snake's nighttime hunts can make sleeping humans a target for snakebites, especially if they sleep on the floor.

Red Spitting Cobra

This red-orange colored cobra uses a superpower to stop enemies in their tracks: spitting venom! The snake squeezes muscles around its venom gland, which is like the organ we use to drool and spit. Pressure builds up like a cork in a bottle, until a small stream of venom shoots out through a small hole in the front of its fangs. These snakes aim for the eyes of their foes, and rarely miss.

Inland Taipan

If someone held a contest for the world's deadliest snake, the inland taipan would win first place. The inland taipan is also known as the "fierce snake," and according to experts, a single bite could have enough venom to kill 100 adults. Thankfully, people rarely come across these snakes in the wild since they are active early in the morning, and they tend to hide in cracks in dry soil and other hard-to-reach places.

Common Death Adder

This snake has a triangular head, short, thick body, and a thin tail. Found throughout Australia, the common death adder is willing to wait for its next meal. It hides itself in leaves, then lies down in a coiled position, twitching its yellow tail to lure in prey. When an animal comes close, the snake quickly strikes and feasts when the prey stops moving.

Tiger Snake

Tiger snakes are found throughout southeastern and southwestern Australia. While they are usually shy, when frightened, they will put up a fight by puffing out their necks with a loud hiss, followed by a venomous bite. Many people in Australia know to steer clear of these snakes when they see them, so they can avoid getting bitten.

SERPENT STORIES:

Studying Snakes

How do we know so much about snakes? By learning from herpetologists (HER-pe-to-le-jists)! These scientists study snakes and other reptiles, and research where snakes live, what they look like, if they are aggressive or shy, what they eat, and how they catch their food. Some experts also study places where snakes and people live near each other, which tends to happen when people use more land to build houses. This can be a big problem in countries where snakebites happen often. When scientists can teach people about the kinds of snakes in their area and how to stay safe around them, this is a win-win for both snakes and humans.

Using Venom as Medicine

Believe it or not, snake venom is not only a deadly weapon for snakes to use against food and foes. The same toxic chemicals that quickly make a mouse go limp can also be made into life-saving medicines to treat things like high blood pressure, health issues related to diabetes, and heart disease. So far, about a dozen drugs have been approved that doctors can use to treat some diseases, but there are around 20 million venom toxins that need to be studied.

A Viper That Saves Lives

According to Dr. Zoltán Takács, a scientist who studies animal venoms, the Jararaca pit viper snake has saved more human lives than any other animal in our history. The venom from this brown-and-tan striped Brazilian viper was used to make a special kind of medicine that can lower blood pressure and prevent people from having heart attacks.

VICIOUS VIPERS:

Frightening Fangs and Shiny Scales

Bright colors and patterns, spiky scales, and dangerous venom—these are all ways to describe bush vipers. These beautiful and feisty snakes can mostly be found in tropical forests in western and central Africa.

African Bush Viper

Even if this colorful snake can't fly or breathe fire, it still looks downright mythical! The African bush viper is also known as a "green bush viper," but it can be found in shades of red, orange, gray, yellow, black, blue, and brown. This makes these cool snakes what scientists call *polymorphic* (pa-lee-MOR-fik). This means that a single species of animal can come in different patterns and colors. They can also change colors as they grow up. These scaly, dragon-like snakes are ambush hunters, so they usually surprise their prey with a sneak attack while hanging upside down from a tree limb.

More Vipers

Like the elapid snakes, all vipers are venomous and have hollow fangs that they use to hunt prey. But there are a few key differences: vipers can fold their fangs into the roofs of their mouths when they aren't using them, they tend to have wide heads shaped like a triangle, and one of their subfamilies, called pit vipers, have special organs on their faces that can sense heat. Neat, huh?

Eyelash Pit Viper

The eyelash pit viper (also known as the "eyelash palm pit viper") has scales above its eyes that look like eyelashes. Found mainly in tropical forests throughout Central America and northern South America, this snake comes in many bright colors such as yellow, "mossy green," or "Christmas" (green with red blotches). They can hide among fruit and flowers to catch prey.

Gaboon Viper

This odd-looking viper has two large scales on its nose that look like a pair of horns. It's also one of the largest and heaviest vipers in the world, growing up to 6 ft. (1.8 m) long and weighing in at up to 25 lbs. (11.3 kg). They also have the longest fangs of any venomous snake at 2 in. (5 cm) long, or about as long as a rubber eraser. These snakes blend in with dead leaves on the forest floor, while they wait for prey to come close.

Russell's Viper

This snake and its deadly bite makes them another member of the "big four" group. Russell's viper is one of the most common venomous snakes in Southeast Asia. Before striking, the snake moves in a series of loops, raises part of its body off the ground, and lets out a loud hiss.

Eastern Copperhead

Eastern copperheads are to blame for more venomous snakebites than any other species in the United States. They are very hard to spot under a pile of leaves or in a hollowed-out log in the forest. They will not bite people if they are left alone, and their bites are rarely deadly. They help protect people by hunting mice and rats that can damage food and spread disease.

Sri Lankan Pit Viper

This jewel-green snake is one of the most eye-catching snakes in Sri Lanka. Also called the green pit viper, this snake usually waits for its prey in the trees. When an unlucky lizard, bird, tree frog, or mouse comes by, the viper attacks, waits for its prey to stop moving, and then swallows it whole.

White-Lipped Island Pit Viper

This bright blue beauty is both highly venomous and aggressive. White-lipped island pit vipers are found throughout Southeast Asia and Indonesia. Once the pit viper has bitten its prey and the venom takes hold, it uses its fangs to "walk" the prey right into its mouth. Yikes!

SUPER SENSES

Scientists found out that vipers are highly skilled snakes in the way that their bodies evolved, or changed, by this point in time. These changes make them experts at killing prey. For example, when vipers open their lower jaws, their fangs rotate from the roofs of their mouths and straight down. This is the perfect position for getting their venom deep into prey. Similar to other snakes, vipers can squeeze a muscle in their cheeks to control how much venom they use at once.

Heat-Sensing Superpower

Most of the species in the viper family are pit vipers. This means that they have two pit organs found between their eyes and nostrils that give them the ability to find prey in dark or distant spaces. The pit organs allow them to "see" the heat that comes off of warm-blooded prey. That's as if you had a built-in pair of infrared goggles that let you see things moving around in the dark! An added perk is that the pit viper's sense organ helps it find the best hangout spots with the perfect temperature.

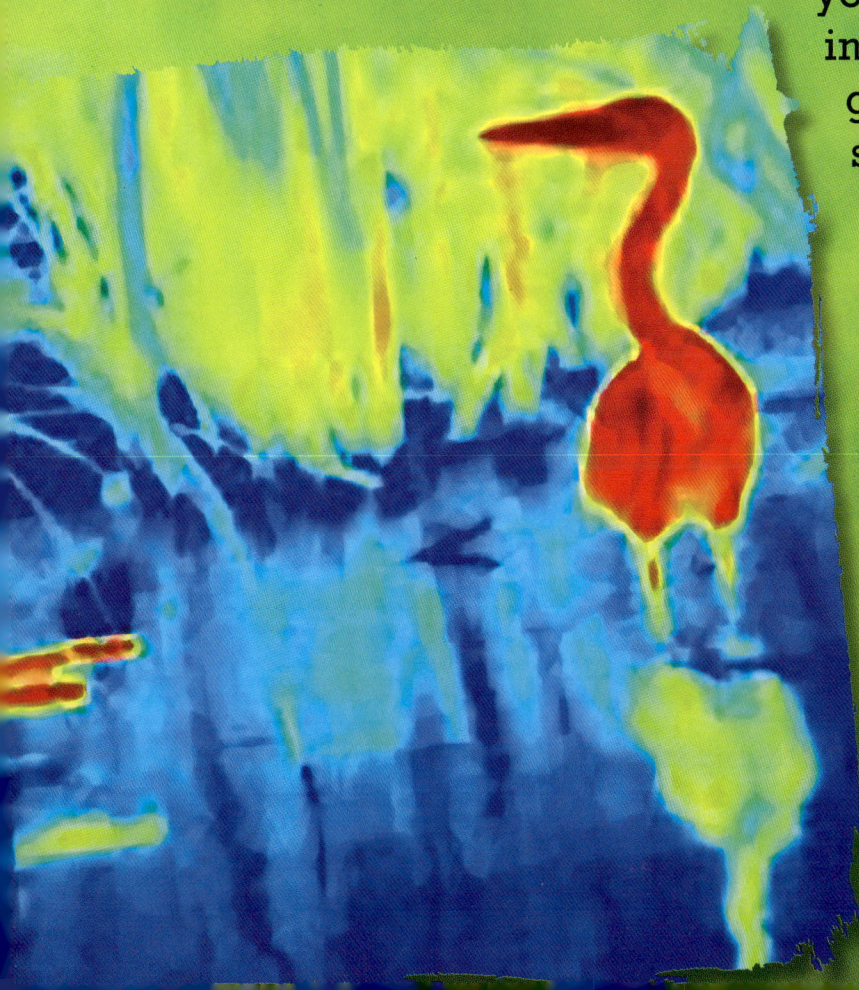

SERPENT STORIES:

The World's Deadliest Island

About 90 mi. (145 km) from the mainland of Brazil, you can find an island that is forbidden to all visitors. Most locals refuse to visit the island. They tell tales about a fisherman and a lighthouse keeper who dared to go, and quickly met their fates. What makes this island so dangerous? Thousands of venomous golden lancehead vipers.

The History of Snake Island

Why is Ilha da Queimada Grande, also called "Snake Island," home to so many deadly vipers? There were rumors that pirates moved the snakes to the island long ago to guard their treasure, but scientists have a more likely explanation. Around 11,000 years ago, sea levels rose and separated the island from the rest of Brazil. In order to survive, their bodies evolved over time to make super-deadly venom that would quickly kill seabirds before they could fly away.

A Dangerous Journey

Despite all the warnings, there are a handful of scientists and government workers who have dared to visit "Snake Island." For those who are willing to make the trip, they'd better watch their step. This is because experts estimate that there is at least one snake per square meter on the island. This means that you'd never be more than 3 ft. (1 m) away from danger! This is why any possible visitors would need to get special permission from the government and have a doctor with them at all times. Even with these safety tips, it's unlikely that "Snake Island" would be on anyone's list of go-to vacation spots.

POWERFUL PYTHONS:

Giant Snakes with a Lethal Grip

Cobras and vipers may be fierce hunters with their long fangs and venom, but you might want to squeeze in some time to read about pythons. Pythons are some of the largest snakes in the world, and they can be found in tropical and subtropical areas in Africa, Southeast Asia, southern China, the East Indies, and Australia. Let's meet some of these not-so-gentle giants!

Reticulated Python

Reticulated pythons are the longest species of snake in the world—they can reach up to 20 ft. (6 m) in length in the wild or almost as long as an average school bus! Reticulated means having a pattern of crossing lines, and this python lives up to its name with repeating black X's on its back that make diamond-like patterns. They move slowly, but what they lack in speed they make up for in hunting skills. Like pit vipers, these snakes have pit organs to sense heat emitted by their prey, so they know just where to strike.

Green Tree Python

These emerald-green snakes are actually born with bright yellow, red, or red-brown scales. Their colors help them hide at different ages. When they're young, the red or yellow colors blend in with forest gaps or edges. But when they are six months to a year old, they slowly turn green and blend in with the leaves and treetops. As adults, they spend most of their time coiled up in trees. To lure prey, the green tree python will wiggle its tail until the curious animal gets close enough. Then, the snake holds on to the branch and strikes.

Even though they're not related, or live near each other, green tree pythons look and act a lot like emerald tree boas. These snakes are so oddly similar that scientists have a name for this situation: parallel evolution. This happens when two different animal groups in different places evolve to look and act alike. In the case of these two snakes, scientists think this happened because they both live in tropical rain forests, and they live up in the trees.

Burmese Python

With their cool giraffe-like patterns and the ability to grow very big, very fast, Burmese pythons are large and in charge. In terms of size, these giants can reach up to 23 ft. (7 m.) long, or the length of a small school bus, and can weigh up to 200 lbs. (90.7 kg)! Instead of fangs or venom, they use their strong bodies to constrict prey. They first find prey using special sensors in their tongues and jaws. Then, they chomp down to hold the animal in place, wind their bodies around it, and squeeze it until the prey dies. To enjoy its meal, the Burmese python separates its jaws and swallows the prey whole.

If you've ever seen or held a big yellow and white snake at a zoo or wildlife center, it could have been an albino Burmese python! These pythons are usually brown, tan, and black, but the albino version is missing special color cells called *pigments*. This makes them white with yellow patches. These snakes have a tough time blending in when they're in the wild, but they are a popular pick for zoos with their beautiful colors and laid-back attitude.

SERPENT STORIES:
Florida's Python Problem

Not-So-Great Pets

After learning about Burmese pythons, it's not hard to see why someone might want to own one as a pet. The problem is that these snakes can grow as long as a bus and as wide as a telephone pole! When a pet owner's snake has gotten too big to handle, they often release it into their backyard. This might not seem like a big deal, but it's very bad news for nearby animals.

The Problem with Pythons

One area where this has become a huge problem is in Florida, USA. These snakes are from southern and southeastern Asia. People either released their pet pythons or they escaped from buildings destroyed in tropical storms and hurricanes. This allowed the snakes to become an invasive species, meaning that they can grow and live in places they shouldn't be. Burmese pythons have been in Florida since the 1980s, but now tens of thousands of them are found in southern Florida and the Everglades National Park. Scientists discovered the python problem when they found the populations of opossums, bobcats, rabbits, and other small mammals were decreasing.

Snake vs. Alligator

Raccoons and rabbits aren't the only animals in Florida that have to watch their backs. These snakes can eat so many different animals that they also chow down on birds, deer, and even alligators! Burmese pythons don't have many predators, so they often face off against alligators for food and space.

BIG BAD BOAS:

So Strong, They're Crushing It

If you thought pythons were cool, just wait until you read about boas. These rad reptiles are found around the world in lots of different places, including woodlands, semi-desert regions, jungles, and forests. Like pythons, boas are constrictors, meaning they grab and hold on tight to their prey until it stops moving. They also come in all shapes and sizes, from giant anacondas to slender tree boas.

Boa Constrictor

With its striking patterns and fearsome hunting style, this famous snake may strike fear in those who come across it at zoos or in the wild. The truth is that unless you're a bird, bat, rat, or mongoose, you don't have to worry about becoming this snake's next meal. In fact, boa constrictors should be thanked for hunting rats, mice, and opossums in places where they can spread diseases to people. These snakes are so good at catching rats that they are used to help get rid of them in some South American homes! They can be found in rain forest edges, grasslands, and semi-desert areas from northern Mexico to Central and South America.

Other Beastly Boas

Besides constrictors, there are many other members of the boa family that are extra shiny, speedy, strong, and supersized. Let's meet some more awesome boas!

Brazilian Rainbow Boa

If all the snakes in the world entered a beauty contest, the Brazilian rainbow boa would certainly win grand prize. In addition to the ringlike pattern on their scales, these reddish-brown fashionistas have tiny ridges on their scales that refract light, making their skin shimmer in lots of different colors. They are found in humid woodland forests in the Amazon River basin, in Guyana by the coast, French Guiana, and southern Venezuela.

Green Anaconda

This massive snake is known for being the heaviest in the world. They can grow to weigh up to 400 lbs. (181.4 kg). That's almost twice as heavy as an average refrigerator! They are found throughout South America and some of the Caribbean Islands, and eat a wide range of animals, including peccaries (a medium-sized wild pig), capybaras, and even jaguars.

Arabian Sand Boa

Unlike many of its rainforest-dwelling family members, the Arabian sand boa makes its home in the deserts of Saudi Arabia. They will burrow deep in sand and soft earth, but then come up to the surface at night to wait for prey. They are different from other boas because their eyes are on the top of their head, instead of on the sides. This makes it easier for the snake to see what's happening on the surface while its body is buried under sand.

Garden Tree Boa

Also called the Amazon tree boa, garden tree boas come in many different colors, from blacks, grays, browns, greens, yellows, oranges, and reds, to a blend of color combinations. As you may have guessed from its name, this snake tends to hang out in trees, especially in the Amazon Rainforest. They can also be found in savannas and dry forests.

Jamaican Yellow Boa

As its name suggests, the Jamaican yellow boa can only be found in the island country of Jamaica. They are nocturnal, which means they sleep during the day, are active at night, and spend most of their time hanging from tree branches or near cave entrances. Young Jamaican boas hunt small lizards and frogs, but the adults go for prey with wings. They're able to snatch birds right out of the air, and can also snack on bats while they fly in and out of their caves.

Rosy Boa

Adult rosy boas can range from 17–44 in. (43–112 cm) long, or about the length of a full-sized guitar. These slender snakes are named after the three dark stripes that run against their light background skin tone. These stripes can be brown or orangish, but they are often a soft rose color. They are found in the southwestern United States and parts of Mexico.

SERPENT STORIES:

A Massive Discovery

The Biggest Snake That Ever Lived

Scientists were shocked when they put together all the pieces of a Titanoboa that they had found over the years. Why? Because they found vertebrae (bones that make up the spine) from multiple snakes that measured between 42–49 ft. (12.8–15 m) long. That means that Titanoboa was as long as a school bus and weighed as much as a small rhino! The ancient snake was likely related to modern boas, but it lived like the anaconda, meaning that it liked to swim in rivers and swamps, and could eat crocodiles.

Titanoboa

In 2002, scientists visited Cerrejón, a very large open coal mine in northern Colombia, South America. This mine also happens to be one of the most important fossil dig sites in the world, and the scientists were not disappointed by what they found. In 2011, they discovered a large fossil that turned out to be an entire skull of a giant snake.

Turn Up the Heat

So how did this monstrous snake get so big? Some experts think this has to do with the hot temperatures in its habitat, or where the snake lived. Reptiles tend to grow bigger in warmer places, where they can take in enough energy from heat to survive. Scientists discovered that the average temperature in that area at the time was at least 82.5 degrees F. (28° C). This giant snake enjoyed living in the tropical heat. If Earth continues to get warmer over the years, we could see more oversized snakes and reptiles someday in the future!

REAL-LIFE SEA MONSTERS:
Dazzling but Deadly Sea Snakes

Old legends speak of sea serpents, dragon-like fish, and monsters from the deep. While they don't really exist, there are a handful of real animals that could have inspired these tales. Some of these so-called sea monsters are actually sea snakes. While they aren't big enough to take down a ship, they are pretty amazing in other ways.

Banded Sea Krait

This beautiful striped sea snake has some of the most toxic venom in the world—ten times more toxic than that of a rattlesnake! Thankfully, snakebites are very rare since this snake is shy, calm, and unlikely to bite people. These snakes are experts at hunting eels in coral reefs, and they can stay safe from predators by moving their tails like they are a second head.

Sea Snakes and Kraits

There are two subfamilies of snakes that spend most of their time in the ocean: sea snakes and sea kraits. Both of these types of sea snakes are actually elapid snakes (in the same family as cobras), but have evolved over time to live in and near the sea. Most sea snakes are found on coral reefs in warm waters around northern Australia, New Guinea, Indonesia, the Philippines, and Southeast Asia. Sea kraits are found in coastal waters in southern Asia, Southeast Asia, Melanesia, and Polynesia. Like their cobra cousins, these aquatic animals are highly venomous.

Yellow-Bellied Sea Snake

This strange-looking blue and yellow snake is built for life at sea. They never need to slither on land, and they will only wash up on shore if they are sick or injured. These sea snakes swim by moving their bodies from side to side under water, and they can also move backward and forward. Its tail is flat and broad, making an excellent paddle to help it swim. Even though they need to come up to the surface to breathe air, they can spend up to ninety minutes or longer under water at a time. They feed on fish by swiping at them sidewise with a venomous bite.

Long-Distance Swimmer

The yellow-bellied sea snake is the only sea snake that spends most of its time away from the shore and coral reefs. They can be found in areas ranging from the eastern coast of Africa to the western coast of Central America and in the Indian and Pacific Oceans. This sea snake is called a *pelagic species*, which means that it lives in the open ocean. It can travel long distances by taking a ride on ocean currents.

Olive-Brown Sea Snake

This sea snake lives up to its name with its olive-brown, golden brown, or gray body, but it often has a pale, cream-colored belly. They're found in the waters of the coral reefs and along the coasts in northern Australia, as well as the southwest Pacific Ocean. They are built for aquatic living, but need to come up for air every once in a while. To help them survive, they have a flat, paddle-like tail and one big lung that lets them stay under water for up to two hours at a time.

Now You Don't See Me

This snake has a cool trick that helps it stay hidden during the day. They have special organs in the skin of their tails that can sense light around them. They use this built-in light sensor to find the best hiding places in tiny cracks and holes in coral reefs. This makes it so the snake is nearly invisible in the daytime, and it can stay safe until it comes out to hunt at night.

SERPENT STORIES:
Real-Life Snake Hunters

Dr. Zoltán Takács:
The Venom Seeker

Dr. Zoltán Takács, scientist and inventor, travels around the world to collect DNA and venom samples from snakes. He explored more than 150 countries, through dense jungles, harsh deserts, and tropical coral reefs. When he returns to his lab, he studies the venom and DNA to find out if it could be used for a new lifesaving medicine. This work is exciting but dangerous, and Takács has been bitten more than a handful of times. But thanks to his lifelong love of nature and adventure, a few injuries hardly get in his way.

Vava Suresh:
The Cobra Whisperer

When it comes to catching cobras, Vava Suresh is the G.O.A.T.! In his home state of Kerala, India, highly venomous snakes—like cobras and vipers—will often wander into people's homes and yards. This is dangerous for both the people and the snakes. Luckily, the people in Kerala know just who to call: the snake whisperer, Vava Suresh! Once he is able to catch the snake and safely move it away from the area, he gives it to the local forest department to release back into the wild. As of 2020, Suresh has rescued nearly 50,000 snakes, including 180 king cobras.

Romulus Whitaker:
The Snake Protector

Romulus Whitaker is known for creating the Chennai Snake Park and the Madras Crocodile Bank Trust in India. He grew up in northern New York State as a child, where he was fascinated by bugs, snakes, and other creepy-crawly critters. His lifelong love of snakes really started when he and his family moved to Mumbai, India. He started working to protect snakes in the 1970s, with the goal of setting up a snake park to teach people about venomous snakes. Nowadays, Whitaker has turned to research and making movies to help others learn about and protect snakes and other reptiles.

Romulus Whitaker

OTHER AWESOME SNAKES:

See Them to Believe Them

RATTLESNAKES

You might have thought you've already learned about all the coolest and deadliest snakes in the world, but we've only scratched the surface. It would be a misss-take for us to not talk about rattlesnakes, some of the most feared predators in North America. Rattlesnakes are also members of the viper family.

Sidewinder

What's that slithering across the sand at record speeds? It's the sidewinder! With their special sidewinding moves across the hot sand, they can reach speeds of up to 18 mph (29 kph). The sidewinder's name and ability are shared by three vipers that are distantly related: the sidewinder rattlesnake, the Peringuey's desert adder, and the desert horned viper. Each of these reptiles lives in similar habitats around the world: the deserts in the American Southwest, the Middle East, and Southern Africa.

Eastern Diamondback

This rattlesnake is the biggest venomous snake in North America, and the largest known rattlesnake. They can reach up to 8 ft. (2.4 m) in length and weigh up to 10 lbs. (4.5 kg)! These stunning animals can be easily spotted by the diamond-shaped pattern on its back, and they are well known for their rattle and painful, toxic bite. When they are bothered, the rattle on this snake's tail shakes out a warning before it strikes. Their rattles can often break off, but a new one grows in each time the snake sheds.

FLYING SNAKES

Yes, you read that right—flying snakes are a real thing.

If you talk to the experts though, they would be more likely to call them gliding snakes, since they tend to glide through the air more like a flying squirrel than a true flying animal like a bird or bat. Based on new research from June 2020, experts got a better sense of how these snakes can stay airborne when moving from tree to tree. As it turns out, these snakes flatten their bodies and move in side-by-side wavelike motions and up-and-down bending movements, gliding through the air like tiny dragons.

Paradise Tree Snake

This beautiful polka-dotted snake spends all of its time in trees in many island countries in southeastern Asia. When it is ready to move to another tree, instead of slithering to the ground like other snakes, it will go to the top of its tree, then jump off and glide gracefully to the next tree. Once the slender snake has lifted off, it can cover a distance of about 328 ft. (100 m). That's almost the distance of an entire American football field!

Golden Flying Snake

Also known as the golden tree snake, this flying snake is thought to be mildly venomous. Despite its name, this snake usually has green or greenish-yellow bands that crisscross over a black background. They are found in India, Sri Lanka, and most of Southeast Asia, and they are most often found on large trees in thick forests. These snakes are very shy and tend to escape danger by gliding away to a nearby tree.

TINY BUT MIGHTY

Last but certainly not least, here are two of the world's smallest snakes. They may be small in size, but they still have some of the same features as their larger relatives. Let's meet these very small serpents!

Brahminy Blindsnake

These tiny snakes are often mistaken for earthworms, and apart from a difference in color, it's not hard to see why. These blindsnakes are usually very thin and a shiny silver-gray color. They are not segmented, having rings around their body, like earthworms. These snakes do tend to hang out in similar places, and they are even known as "flowerpot snakes" for their habit of stowing away in the soil of houseplants. This snake can grow to a mere 7 in. (17.8 cm) at most, but they can still cause problems as an invasive species. They come from Southeast Asia, but these little snakes have wormed their way into Florida and California, USA.

Barbados Threadsnake

In August 2008, scientists found the world's smallest snake on the Caribbean Island of Barbados. At full size, this little snake is less than 4 in. (10 cm) long. It is also as thin as a spaghetti noodle, and it can curl up on a US quarter without needing much extra space. The snake eats termites to survive, and they burrow in soil like the blindsnakes, but not much else is known about this tiny living noodle.

Conclusion

Even though you've reached the end of this book, there are still so many other ways to learn about these amazing reptiles.

Visit your local library and pick up other books about snakes, or with help from an adult, research snake facts online. There's also your local zoo or wildlife center, and you can even find snakes in your own backyard or local park! Just remember to give any snake that you see plenty of space (a distance of at least 6 ft. (1.8 m)) and never try to touch or pick one up.

These animals are often misunderstood. Even in places where snakebites aren't a big problem, many people are scared about getting bitten or hurt. Talk to your family and friends about what you learned, your favorite snakes, and how to stay safe around them. Remember, most snakes are more afraid of you than you are of them, and they will only bite if they feel threatened or if you get too close to them. Share what you've learned and stay curious. Maybe one day you will be the next person to turn venom into medicines or find a new snake species! The possibilities are endless.